AMAZING MILITARY MACHINES

FIGHTER PLANES

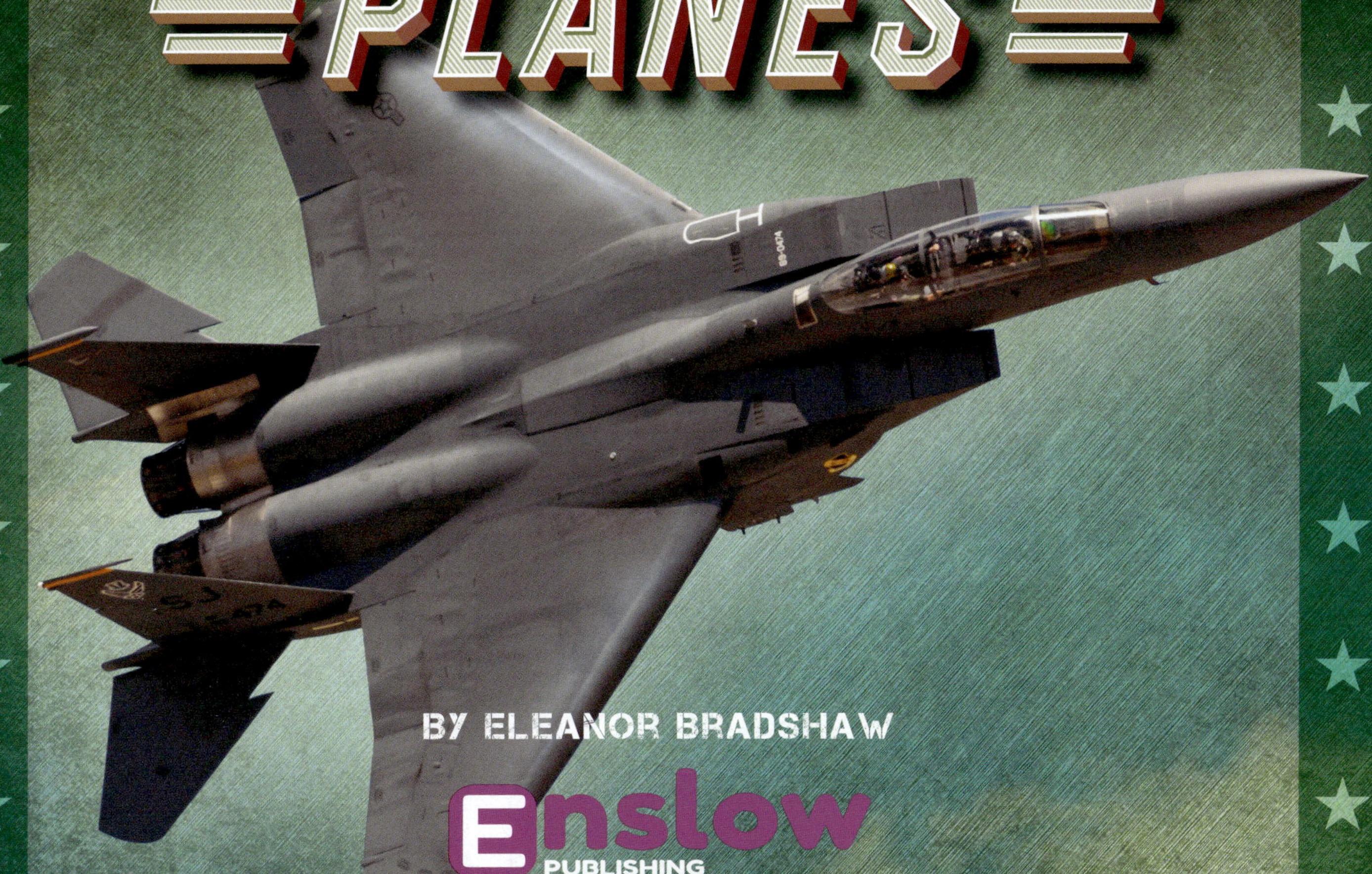

BY ELEANOR BRADSHAW

Enslow PUBLISHING

Please visit our website, www.enslow.com. For a free color catalog of all our high-quality books, call toll free 1-800-398-2504 or fax 1-877-980-4454.

Library of Congress Cataloging-in-Publication Data
Names: Bradshaw, Eleanor, author.
Title: Fighter planes / Eleanor Bradshaw.
Description: Buffalo, New York : Enslow Publishing, [2025] | Series: Amazing military machines | Includes bibliographical references and index.
Identifiers: LCCN 2024000750 | ISBN 9781978539624 (library binding) | ISBN 9781978539617 (paperback) | ISBN 9781978539631 (ebook)
Subjects: LCSH: Fighter planes–United States–Juvenile literature.
Classification: LCC UG1242.F5 B732 2025 | DDC 623.74/640973–dc23/eng/20240206
LC record available at https://lccn.loc.gov/2024000750

Published in 2025 by
Enslow Publishing
2544 Clinton Street
Buffalo, NY 14224

Portions of this work were originally authored by Ryan Nagelhout and published as *Fighter Planes*. All new material in this edition is authored by Eleanor Bradshaw.

Designer: Andrea Davison-Bartolotta
Editor: Rachael Morlock

Photo credits: Cover Eliyahu Yosef Parypa/Shutterstock.com; series art (metal background texture) caesart/Shutterstock.com; series art (title font) TortugaStudios/Shutterstock.com; p. 5 (bottom) Photo by Rusty Baker/U.S. Navy/DVIDS; p. 5 (top) U.S. Air Force photo by R.J. Oriez/DVIDS; p. 7 courtesy of Library of Congress; p. 9 File:Downward view of a "Chateau Thierre Aeroplane", a World War I aircraft, in flight over Argone Forest and French trenches, ca.1914-1918 (CHS-11659).jpg/ Wikimedia Commons; p. 10 U.S. Air Force photo/Airman 1st Class Christopher S. Sparks/DVIDS; p. 11 courtesy of NASA/ DVIDS; p. 13 (inset, main) U.S. Air Force Photo by Heide Couch/DVIDS; p. 15 (bottom) Photo by Timothy Sandland/102nd Intelligence Wing/DVIDS; p. 15 (top) File:Fifth Air Force, Korea-Keeping a wary eye open for Russian-built MIG-15s, the pilots of these U.S. Air Force F-86... - NARA - 542259.tif/Wikimedia Commons; p. 16 Photo by Master Sgt. Jeremy Lock/ Defense Imagery Management Operations Center/DVIDS; p. 17 U.S. Air Force photo by Staff Sgt. Staci Miller/RELEASED/ DVIDS; p. 19 MyPlanePics/Shutterstock.com; p. 21 U.S. Air Force photo by Senior Airman Dillian Bamman/DVIDS.

Printed in the United States of America

CPSIA compliance information: Batch #CS25ENS: For further information contact Enslow Publishing, at 1-800-398-2504.

CONTENTS

Boldface words appear in Words to Know.

AIR FORCE PLANES

An air force is a branch of the military that uses airplanes. Some planes move people and supplies from place to place. Some planes, called bombers, **attack** things on the ground. Other planes attack flying aircraft. These are fighter planes!

The U.S. Air Force has a large **fleet** of fighter planes ready for battle.

FIRST FLIGHT

Orville and Wilbur Wright flew the first powered airplane in 1903. Early airplanes used **propellers** to move forward and wings to help the plane rise off the ground. In 1909, the Wright brothers sold the world's first military airplane to the U.S. Army.

Orville Wright was the **pilot** on the first flight of the Wright Flyer at Kitty Hawk, North Carolina, on December 17, 1903.

WORLD WAR I

The United States, among other countries, used planes to fight in World War I (1914–1918). One of the best fighter planes was the French SPAD that held one pilot. The Fokker triplane, which had three wings, was used by Germany.

The SPAD was a biplane, which means it had two wings lined up one above the other.

WORLD WAR II

Air **combat** changed in World War II (1939–1945) with new fighter planes. Small, fast, all-metal airplanes would shoot down enemy bombers to stop attacks. Battles also took place in the air between fighter planes. These were called "dogfights."

P-47 THUNDERBOLT

In World War II, the United States used planes such as the P-51 Mustang, shown here, and the P-47 Thunderbolt.

NEW POSSIBILITIES

The jet **engine** changed flight and opened new possibilities for fighter planes. Jet engines work by taking in a large amount of air and turning it into a high-speed jet of gas. This pushes the plane forward and makes it go very fast.

Jet engines are often set beneath the wings of a plane.

JET FIGHTERS

The first major use of U.S. jet fighters was in the Korean War (1950–1953). As jets flew faster, pilots also learned to avoid being spotted by enemy **radar**. Many new jets started to use **stealth technology** in the 1980s and 1990s.

Americans flew F-86 Sabres, shown here, to battle Russian-built fighter planes during the Korean War.

A jet fighter called the F-15 was first used in 1972. It has two jet engines and room for two pilots in the **cockpit**. Another powerful fighter is the F-22 Raptor, which came into use in 2005. Two engines bring the F-22 to top speeds.

Called the "Strike Eagle," the F-15 can attack targets in the air and on the ground.

LATEST AND GREATEST

The F-35 is the latest jet to join the U.S. fighter fleet. It has one engine, seats one pilot, and can take off straight up in the air! It has advanced stealth technology and can be used by different branches of the military.

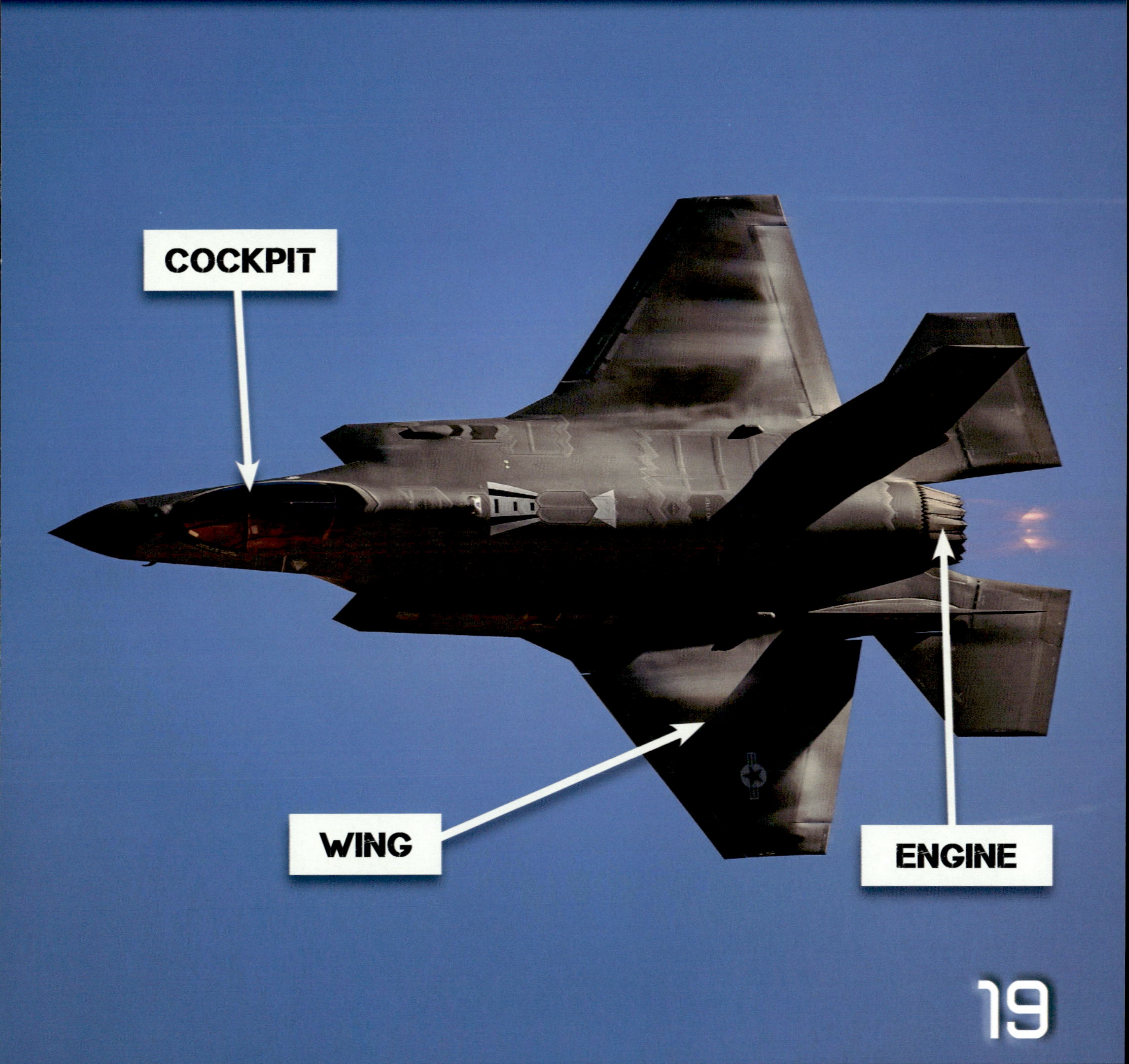
COCKPIT
WING
ENGINE

ONWARD AND UPWARD

Today, fighter planes do more than battle other planes in the sky. The U.S. military is working on another fighter plane that can serve many purposes. It will be able to fight other planes, carry out ground attacks, and use advanced **weapons**.

Some fighter jets escort, or give guided protecton, to other planes.

WORDS TO KNOW

attack: To set upon something with force.

cockpit: The place a pilot sits in a plane.

combat: Armed fighting between opposing forces.

engine: A machine that makes power.

fleet: A group of planes under one command.

pilot: A person who flies a plane.

propeller: Paddle-like parts on a plane that spin to move the plane forward.

radar: A machine that uses radio waves to locate and identify objects.

stealth technology: The tools used to keep something from being seen or spotted.

weapon: Something used to cause someone or something injury or death.

FOR MORE INFORMATION

BOOKS

Cella, Clara. *Fighter Pilots.* Minneapolis, Minnesota: Lerner Publications, 2023.

Gitlin, Marty. *Fighter Jets.* Hallandale, Florida: Mitchell Lane Publishers, 2021.

Ransom, Candice. *How Fighter Jets Work.* Minneapolis, Minnesota: Lerner Publications, 2020.

WEBSITES

Aviation
airandspace.si.edu/explore/topics/aviation
Learn about the history of flight on this page from the Air and Space Museum.

F-35A Lightning II
www.af.mil/About-Us/Fact-Sheets/Display/Article/478441/f-35a-lightning-ii/
Find out about the F-35 in this fact sheet from the U.S. Air Force.

Publisher's note to educators and parents: Our editors have carefully reviewed these websites to ensure that they are suitable for students. Many websites change frequently, however, and we cannot guarantee that a site's future contents will continue to meet our high standards of quality and educational value. Be advised that students should be closely supervised whenever they access the internet.

INDEX